Wind
What Can It Do?

by Janet McDonnell
illustrated by Gwen Connelly

Created by

Distributed by CHILDRENS PRESS®
Chicago, Illinois

Grateful appreciation is expressed to Elizabeth Hammerman, Ed. D., Science Education Specialist, for her services as consultant.

CHILDRENS PRESS HARDCOVER EDITION
ISBN 0-516-08123-3

CHILDRENS PRESS PAPERBACK EDITION
ISBN 0-516-48123-1

Library of Congress Cataloging in Publication Data

McDonnell, Janet, 1962-
Wind : what can it do? / by Janet McDonnell ; illustrated by Gwen Connelly.
p. cm. — (Discovery world)
Summary: Describes what wind is and what it can do, from pushing clouds and pulling kites to making giant waves. Includes instructions for making a windsock.
ISBN 0-89565-555-1
1. Winds—Juvenile literature. [1. Winds.] I. Connelly, Gwen, ill. II. Title. III. Series.
QC931.4.M4 1990
551.5'18—dc20
89-24011
CIP
AC

1 2 3 4 5 6 7 8 9 10 11 12 R 99 98 97 96 95 94 93 92 91 90

Wind
What Can It Do?

Where is Discovery World?
It can be anywhere—
anywhere at all.
It can be at a zoo,
a museum, or beside the sea.
It can be in a pond,
a library, or up in a tree.
All you need is curiosity.
It can be right at
your own back door.
All you need to do is EXPLORE!

WORLD
and friends
N
W
E
S
So come along and find out more about . . .
WIND!

What can wind do? So many things. Just look around you!

Sometimes wind can push you around. It can catch your cap and send it sailing.

It can pull your kite and take it right up to the clouds. You can feel wind tug on the string.

Sometimes wind can push clouds around. It can push them into shapes that look like animals. Or it can blow them all away.

Sometimes wind can be very strong.
Tornado winds can lift trees and cars, . . .

and hurricane winds can make giant waves!

What can wind do? It can blow cold from the north and make your ears sting, . . .

or it can blow warm from the south and
melt the snow.

It can cool you down on a hot, summer day or give you goose bumps when you get out of a pool.

What can wind do? It can scatter seeds to new places to grow.

And it can carry smells through the air to your nose.

It can turn windmills so they pump
water up from underground, . . .

and it can dry your clothes when they
hang on a line.

What can wind do? It can give birds a ride so they can rest their wings.

And sometimes wind can even make noise. It can whistle through windows . . .

and whisper through trees.

Wind can do all of these things because wind is air that moves.

NOW EXPLORE SOME MORE WITH PROFESSOR FACTO!

Wind can be very helpful in many ways, especially when it comes to moving people around! Here are some ways that people "ride the wind"!

gliding

windsurfing

ballooning

sailboating

You can see windpower in action with a pinwheel. It is simple to make. Here's how:

1. Cut out a square like the one below from construction paper.
2. Cut from the corners to the small square in the center.
3. Bring the marked corners to the dot in the center and tape them down. Bend but do not fold the paper.
4. With the help of an adult, press a pushpin through the center of the pinwheel. Then press the pin and the pinwheel into the wood at the top of a pencil. Do not push the pin in too far or the pinwheel will not spin.

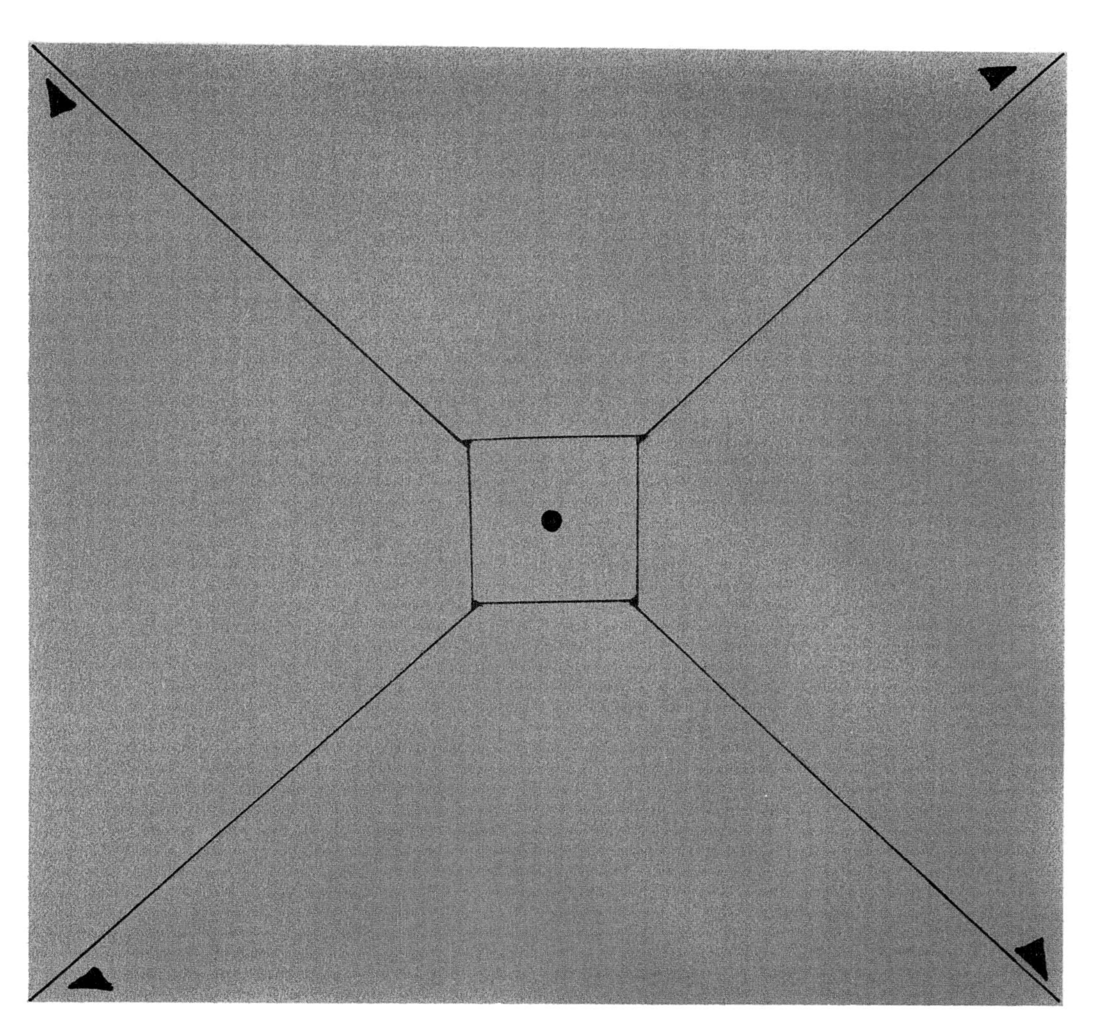

How can you tell which way the wind is blowing? Well, you can look to see which way the leaves blow across the ground, or you can see which way your hair blows, or you can make a WINDSOCK!

Here's what you will need:

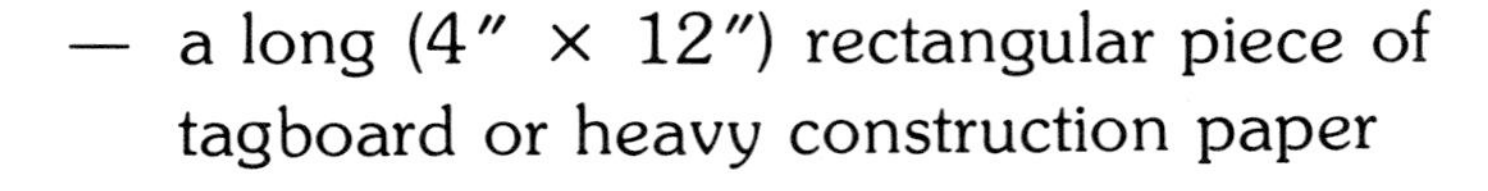

— a long (4″ × 12″) rectangular piece of tagboard or heavy construction paper

— string or yarn

— crayons or markers

— a paper clip

— 5-10 strips of cloth, tissue paper, or crepe paper

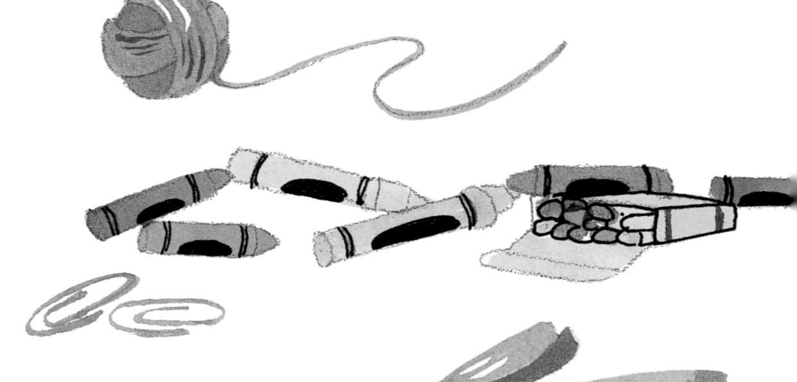

— an adult with a stapler and a hole puncher

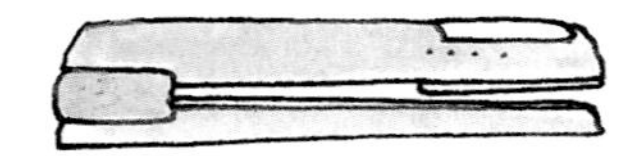

1. First decorate the tagboard with markers or crayons.
2. Have an adult staple the long edges of the tagboard together to make a big tube.
3. Have an adult punch three holes (equal distances apart) around the top of the tube.
4. Tie a 6″ piece of string through each hole.
5. Pull the ends of the three strings together and tie them onto the paper clip.
6. Have an adult staple the strips of cloth to the other end of the tube.

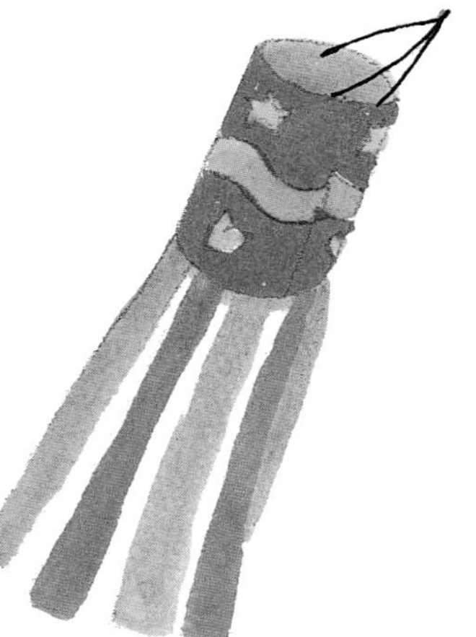

Now you've made a windsock! You can tie it to a tree branch to tell which way the wind is blowing.

6/22

INDEX